Kelen Schelbauer

Diversity of Lepidoptera in APP BR-116 from Mafra to Papanduva, SC

Kelen Schelbauer

Diversity of Lepidoptera in APP BR-116 from Mafra to Papanduva, SC

Butterflies and moths found in the region

ScienciaScripts

Cover image: www.ingimage.com

This book is a translation from the original published under ISBN 978-3-330-77366-0.

Publisher:
Sciencia Scripts
is a trademark of
Dodo Books Indian Ocean Ltd. and OmniScriptum S.R.L publishing group

120 High Road, East Finchley, London, N2 9ED, United Kingdom
Str. Armeneasca 28/1, office 1, Chisinau MD-2012, Republic of Moldova, Europe
Printed at: see last page
ISBN: 978-620-7-27921-0

ACKNOWLEDGEMENTS

Several people have been responsible for getting me this far. In view of this, I could not fail to mention them with great affection, love and friendship.

I thank God first and foremost for always being in front of me, guiding and protecting me with divine light.

Thank you, Mum, for helping me on this journey, for believing in me since I was a child and encouraging my studies, for financing me, patiently putting up with difficult times and making this moment possible. To my sisters (Helen and Renata), brother-in-law (Leandro) for their support and help with my work, and especially for helping me catch butterflies in the field; to little Louis for his games and for always waking me up in the morning.

To Cristiano, for your affection and support in times of despair, thank you for helping me with my work and for keeping me company all these years. I'll be helping you with whatever you need too, love.

I would like to thank all the teachers for their teachings, especially my advisor Professor Mário Fritsch for the information he passed on, for the "tugs on the ear", for encouraging my admiration for butterflies so that it was possible to carry out this enjoyable work; and who once said: "Do something you love, and you'll never have to work again in your life". Who contributed enormously to my growth.

To all my friends, especially my friend Susiele, for always helping me, listening to me and drawing my attention when I was wrong, no matter how much I didn't want to believe it. Thanks also to all my classmates for sharing their anguish and experiences.

And life for the simple fact of existing!

Thank you all! Please know that each and every one of you has played an important role in making me grow, evolve and get this far.

"There will be no butterflies if life doesn't go through long, silent metamorphoses."

(Rubens Alves)

SUMMARY

The order Lepidoptera is a good tool for conservation biology, as it is sensitive to environmental variations. They also play an important role in many pollination processes and make up important trophic levels in the food chain. They are important in ecosystems due to their interactions with vegetation and fauna. The aim of this study was to determine the families of butterflies (Lepidoptera) in Permanent Preservation Areas (APPs) along the BR-116 motorway, from Mafra to Papanduva, SC. Collections took place from March to September, with a sampling effort of 27 hours. A total of 138 individuals belonging to ten families were recorded. It was observed that APP km 031 + 815 showed the greatest diversity, with seven families (Acracidae, Danaidae, Heliconidae, Morphoidae, Nymphalidae, Pieridae, Satyridae). The APPs km 026 + 780 and km 061 + 925 had the lowest diversity, with only one family recorded (Saturidae). In the APPs surveyed there were butterflies from the families Acracidae, Danaidae, Hesperiidae, Heliconidae, Lycaenidae, Morphoidae, Nymphalidae, Pieridae, Riodinidae and Satyridae.

Keywords: Diversity. Lepidoptera. Permanent Preservation Area - APP. Pollination.

SUMMARY

CHAPTER 1

INTRODUCTION

1.1 PRESENTATION OF THE THEME

Insects of the lepidoptera order attract attention because of their colour and variety, which makes them very interesting and admirable to people. Butterflies also have a multitude of meanings related to the soul, transformation, freedom and death in ancient cultures.

They are characterised by the scales on their wings, which gave rise to the order's name. Species of the order Lepidoptera are very important for diagnosing environmental quality because they are easy to identify in nature and have a rapid life cycle, making them very susceptible to change. In addition, they play a fundamental role in ecological interactions, in the food chain, and in pollination through the exchange of genetic material.

It should be noted that to date there has been no study of the diversity of lepidopterans found in the proposed study area, so it was decided to carry out this research to provide subsidies for scientific knowledge.

1.2 PROBLEM

Which families of Lepidoptera (butterflies) are found in Permanent Preservation Areas (PPAs) along the BR-116 motorway, from Mafra to Papanduva, SC?

1.3 BACKGROUND

The purpose of this study was to survey the families of Lepidoptera (butterflies) that occur in Permanent Preservation Areas (PPAs) along the BR-116 motorway, from Mafra to Papanduva, SC.

The order Lepidoptera is widely used in studies and surveys of the conservation status of forests, as they are very sensitive to environmental variations, as well as being easy to identify in the environment, abundant and with a rapid life cycle, so monitoring lepidoptera becomes a very important tool in Conservation Biology.

Lepidoptera have great ecological importance in many pollination processes and make up important trophic levels in the food chain. Flowers have a set of characteristics that condition them to be used more frequently by one or other potential pollinating agent: these are called pollination syndromes.

Pollination is the process of transferring pollen from the anthers of a flower to the stigma, which can be from the same flower (self-pollination), which can take place without the need for a pollinating agent or with its participation, or from another flower of the same species (cross-pollination), always with the participation of a pollinating agent, such as wind, birds, insects or another. The pollen grain fertilises the ovule to form the seed, while the ovary develops to form the fruit (EMBRAPA, 2013). Cross-pollination involves the exchange of genetic material between species, meaning that no two organisms are the same, making them more resistant to environmental variations.

Certain species of Lepidoptera depend exclusively on one plant species, which makes them sensitive to variations in the environment; they can therefore be considered indicator species of environmental quality.

The importance of this work is the contribution it will make to verifying the condition of vegetation regeneration in the Permanent Preservation Areas (PPAs) surrounding the motorway, taking into account that lepidopterans are important in tropical ecosystems due to their interactions with vegetation (GILBERT, 1984 *apud* RAFAEL *et al.,* 2012). However, knowledge of the biology of these insects can help us understand countless processes (such as pollination) that directly or indirectly maintain the survival of many other organisms, including the human species (RAFAEL *et al*., 2012).

1.4 OBJECTIVES

1.4.1 General Objective

To determine the families of Lepidoptera (butterflies) in Permanent Preservation Areas (APPs) in the right-of-way of the BR-116 motorway, from Mafra to Papanduva, SC.

1.4.2 Specific objectives

- Determine the families of lepidopterans (butterflies) found in the study area; -

Demonstrate the diversity of lepidopterans (butterflies) found along the motorway's right of way;

- To provide information on the entomofauna of lepidopterans (butterflies) in the region;
- Contribute to the monitoring and conservation of the APPs along the stretch through information on the pollinating agents that occur in the areas.

CHAPTER 2

THEORETICAL FRAMEWORK

2.1 AUTOPIST

Since 2008, Autopista Planalto Sul has been the company responsible for the 412.7 kilometres of federal highway BR-116, which connects the city of Curitiba/PR to the border between the states of Santa Catarina and Rio Grande do Sul. The motorway was built between the 1940s and 1950s.

The concession to manage the stretch was obtained through an auction held on 9 October 2007, which lasts 25 years.

The route cuts through the municipalities of Fazenda Rio Grande, Mandirituba, Quitandinha, Campo do Tenente and Rio Negro, in the state of Paraná. Mafra, Itaiópolis, Papanduva, Monte Castelo, Santa Cecília, Ponte Alta do Norte, São Cristóvão do Sul, Ponte Alta, Correia Pinto, Lages and Capão Alto, in the state of Santa Catarina.

2.2 DIVERSITY OF LEPIDOPTERA (BUTTERFLIES) IN PERMANENT PRESERVATION AREAS

2.2.1 Phylum Arthropoda

The main characteristic of the insects of the Arthropoda Phylum is their articulated appendages, from the Greek, *arthron* = articulation + *podos* = feet, one of the most populous with around 1,000,000 known species, making it the largest phylum in the animal kingdom.

They are triblastic, protostomous and coelomate animals. Apart from annelids, they are the only metamericly segmented invertebrates, but they don't have intersegmental septa. During embryonic development, some metameres may fuse, making it difficult to visualise the segments. With
Bilateral symmetry, the body is organised into head, thorax and abdomen or cephalothorax and abdomen, with six legs distributed in three pairs, attached to the thorax.

They have a complete digestive system, with jaws adapted for chewing or sucking. They have a lacunar or open-type circulatory system where the heart is dorsal and pumps the haemolymph to the anterior end, reaching the body lacunae or haemoceles, the return of the haemolymph occurs through lateral orifices called ostia.

Respiration through the body surface, gills, tracheae (air tubes), or foliaceous lungs. Excretory system with paired excretory glands (coxal, antennal or maxillary glands), some with Malpighian tubules.

Nervous system formed by a dorsal ganglion connected by a ring that surrounds the digestive tract to a double ventral nerve chain made up of segmental ganglia; fusion of the ganglia occurs in some species; well-developed sensory organs (HICKMAN *et al.*, 2013).

Separate sexes, with paired reproductive organs and a duct system. With internal fertilisation, oviparous or ovoviviparous, frequent metamorphosis, parthenogenesis in some cases (HICKMAN *et al.*, 2013).

The entire external surface of the body is covered by a chitin exoskeleton secreted by the epidermis. Chitin is a nitrogenous polysaccharide ($C8H18O5N$) that is insoluble in water, alcohol, dilute acids or the digestive juices of other animals. The exoskeleton gives the phylum a restriction on growth, requiring a periodic change for growth, called moulting or ecdysis. The abandoned exoskeleton is called exuvia. Its main success is due to its exoskeleton, which provides protection against predators, parasites, insulates it from the environment and protects it from desiccation.

2.2.1.1 Classification of the Phylum Arthropoda

According to Hickman *et al*, 2013, the phylum Arthropoda is divided into:

- Subphylum 1 - Trilobite: trilobites, from the Cambrian to the Carboniferous period, all extinct; body divided into head, thorax and abdomen and with two longitudinal lines marking three body lobes - the right, left and median.

- Subphylum 2 - Chelicerata: eurypterids, limpets, spiders, ticks. The first pair of appendages modified into chelicerae; one pair of pedipalps and four pairs of legs; no antennae and no mandible, generally the cephalothorax and abdomen not segmented.

I. Class Merostomata - aquatic chelicerates. Cephalothorax and abdomen; lateral compound eyes, appendages with gills; pointed telson.

1. Subclass Eurypteria - giant sea scorpion (extinct).

2. Subclass Xiphosurida - limulus. Example: *Limulus*.

II. Class Pycnogonida - sea spiders. Small, 3 to 4 mm, but some reach 500 mm; body basically reduced to the cephalothorax; tiny abdomen; usually with four pairs of long locomotive legs (some with 5 or 6 pairs); mouth located on a long proboscis; four simple eyes; no respiratory or excretory system. Example: *Pycnogonum*.

III. Class Arachnida - scorpions, spiders, mites, ticks, opilions. Four pairs of legs; abdomen segmented or not, presenting appendages or not, and generally distinct from the cephalothorax; respiration through gills, tracheae or foliaceous lungs; excretion through Malpighian tubules or coxal glands; dorsal bilobed brain connected to a ventral ganglionic mass in nerves, simple eyes; mainly oviparous, without true metamorphosis. Examples: *Argiope*, *Centruroides*.

- Subphylum 3 - Myriapoda: myriapods. Unirremic appendages, appendages consisting of a pair of antennae, a pair of mandibles and two pairs of maxillae.

I. Class Diplopoda - snake lice. Almost cylindrical body; head with short antennae and simple eyes; body with a variable number of somites; short legs, usually two pairs of legs per somite; oviparous. Examples: *Julus*, *Spirobolus*.

IV. Class Chilopoda - centipedes. Body dorsoventrally flattened; variable number of somites, each with a pair of legs; one pair of long antennae; oviparous. Examples: *Cermatia*, *Litbobius*, *Geopbilus*.

V. I. Class Pauropoda - pauropods. They measure from 1 to 1.5 mm; cylindrical body made up of double segments and with 9 or 10 pairs of legs; no eyes. Examples: *Pauropus*, *Allopauropus*.

VI. Class Symphyla - syrphids. Slender, 1 to 8 mm long, with long, filiform antennae; body made up of 15 to 22 segments with 10 to 12 pairs of legs; no eyes. Example: *Scutigerella*.

- Subphylum 4 - Crustacea: crustaceans. Mostly aquatic, with gills, cephalothorax usually with dorsal carapace, birefringent appendages. Cephalic appendages consisting of two pairs of antennae, a pair of mandibles and two pairs of maxillae.

I. Class Remipedia - remipedes. Carapace absent; protopodites with one article; antennules and antennae birremes; all trunk appendages similar; cephalic appendages large and raptorial; maxilliped segment fused to cephalic appendages; trunk not regionalised. Example: *Speleonectes.*

II. Class Cephalocarida - cephalocarids. Carapace absent; filipodia, protopodites uniarticulate; antennulae unirremes and antennae birremes; compound eyes absent; thoracic appendages. Example: *Hutchinsoniella.*

III. Class Branchiopoda - branchiopods. Carapace present or absent; maxillipeds absent; antennules reduced; compound eyes present; abdominal appendages absent; maxilla reduced.
  - Order Anostraca - Artemia. Examples: *Artenia, Branchinecta.*
  - Order Notostraca - Examples: *Triops, Lepidurus.*
  - Order Diplostraca - Water fleas. Examples: *Daphnia, Leptodora, Lynceus.*

IV. Class Ostracoda - ostracodes. Bivalve carapace entirely enveloping the body; body unsegmented or with indistinct segmentation; no more than two pairs of appendages on the trunk. Examples: *Cypris, Cypridina, Gigantocypris.*

V. Class Maxillopoda - maxillopods. Generally with five cephalic segments, six thoracic, four abdominal plus one telson, although reductions are common; no conspicuous appendages on the abdomen; single-structured nauplius eye; carapace present or absent.

Subclass 1: Mystacocarida - carapace absent; body has head and trunk with ten segments; telson with pincer-like branches; cephalic appendages almost identical, but antennae and mandibles birrhine, other cephalic appendages unirrhine; small, uniarticulate appendages from the second to fifth trunk segments. Example: *Derocheilocaris.*

Subclass 2: Copepoda - Copepods. Carapace absent; thorax with seven segments, of which the first, and sometimes the second, is fused with the head to form a cephalothorax; antennules unirremed; antennae unirremed or birremed; four to five pairs of swimming appendages; parasitic forms often much modified. Examples: *Cyclops*, *Diaptomus*, *Calanus*.

Subclass 3: Tantulocarida - unrecognisable cephalic appendages, except for the antennae of the female in the reproductive form; a solid median cephalic stylet; six free thoracic somites, all with a pair of appendages, the anterior five birremes; six abdominal segments; they are diminutive ectoparasites similar to copepods. Examples: *Basipodella*, *Deoterbron*.

Subclass 4: Branchiura - oval body, head and most of the trunk covered by a flattened carapace, incompletely fused with the first thoracic segment; thorax with four pairs of birrhine appendages; bilobed unsegmented abdomen; compound eyes; reduced antennae and antennules; maxillae often form suckers. Examples: *Argulus*, *Chonopeltis*.

Subclass 5: Pentastomida - Pentastomids. Unsegmented vermiform body, with five small anterior protuberances, four with claws and the fifth has a mouth with suckers. Examples: *Linguatula*, *Armillifer*.

Subclass 6: Cirripedia - barnacles. Sessile or parasitic as adults; reduced head and rudimentary abdomen; absent compound paired eyes; body with indistinct segmentation; generally hermaphroditic; in free-living forms, the carapace becomes an organ for attaching to the substrate and then disappears. Examples: *Balanus*, *Policipes*, *Sacculina*.

VI. Class Malacostraca - generally with eight segments on the thorax and six segments plus telson on the abdomen; appendages on all segments; antennules often birrhine; the first of the initial thoracic appendages, or up to three of them, are often maxillipeds; the carapace covers the head and partially or totally the thorax, sometimes absent; gills are generally thoracic epipodites.

- Order Isopoda - isopods.

- Order Amphipoda - amphipods.
- Order Euphausiacea - krill.
- Order Decapoda - shrimps, crabs and lobsters.

- Subphylum 5 - Hexapoda: hexapods, body divided into a distinct head, thorax and abdomen, one pair of antennae, modified mouthparts, head with six fused somites, thorax with three somites, abdomen with a variable number of somites. Thorax usually with two pairs of wings and three pairs of articulated legs, dioecious, usually oviparous, gradual or abrupt metamorphosis.

I. Class Entognatha: the base of the mouthparts is inside the cephalic capsule; the mandibles have a joint.
   - Protura Order - proturos
   - Order Diplura - dipluros.
   - Order Collembola - springtails.

II. Class Insecta - the bases of the mouthparts are exposed, outside the cephalic capsule; the mandibles usually have two articulation regions.

   Subclass 1: Apterygota - most primitive insects.
   - Order Thysanura - moths.

   A. Infraclass Palaeoptera: wings do not fold over the body; articulation by fused axillary plates.
      - Order Ephemeroptera: ephemeral.
      - Order Odonata - dragonflies.

   B. Infraclass Neoptera: fold the wings over the body; articulation by movable sclerites at the base of the wing.
      - Order Orthoptera - grasshoppers, crickets, hoppers.
      - Order Blattodea - cockroaches.
      - Order Phasmatodea - woodworms and leafworms.

      - Order Mantodea - praying mantis.
      - Order Mantophasmatodea - gladiators.
      - Order Desmaptera - scissors.

- Order Plecoptera - plecoptera.
- Order Isoptera - termites.
- Order embiidina - embiids.
- Order Psocoptera - book lice.
- Order Zoraptera - zorapters.
- Order Phthiraptera - lice.
- Order Thysanoptera - lacerdinhas or tripods.
- Order Hemiptera - bedbugs, cicadas, aphids, jequitiranaboia, mealybugs.
- Order Neuroptera - lion ants, sandflies.
- Order Coleoptera - beetles, fireflies, weevils.
- Order Strepsiptera - strepsipterans.
- Order Mecoptera - mechoptera.
- Order Lepidoptera - butterflies and moths.
- Order Diptera - flies and mosquitoes.
- Order Trichoptera - water fly, frigana.
- Order Siphonaptera - fleas.
- Order Hymenoptera - ants, bees and wasps.

2.2.2 Class Insecta

The Insecta (gr. *insectus,* segmented), correspond to more than 1.1 million catalogued species; experts estimate that there may be around 30 million species, there are more insects than all other classes of animals combined (HICKMAN *et al*, 2013).

Insects differ from other groups of animals in that they have ectogynous mouthparts and generally two pairs of wings, although some have one pair and others are primitively or secondarily apterous. Their size can vary from 1 mm to 20 cm. They are the only invertebrates with the ability to fly, making it easier to escape from enemies, capture prey and find mates. Insect wings are formed by projections from the body wall, and fossil evidence suggests that insects may have developed functional wings more than 400 million years ago.

They spread to practically all *habitats*. Because they are relatively small, they can be carried by water and air to other regions. Their eggs, when well protected, can withstand harsh conditions. Their success is due to characteristics such as wings, a waterproof cuticle, mechanisms to minimise water loss and the ability to become dormant during adverse conditions (HICKMAN *et al*, 2013).

Feeding is very varied among insects, and the mouthparts are specialised for each type of food. They breathe through a tracheal system. Excretion takes place through Malpighian tubules. The sexes are separate and fertilisation is generally internal. Almost all of them undergo metamorphosis, which can be hemimetabolous (gradual) where the adults emerge from the last moult of the nymph stage, and holometabolous (complete) metamorphosis in which the last larval moult gives rise to the pupa (a stage that does not feed) and the adult emerges from the last moult carried out by the pupa.

They are responsible for the pollination of all phanerogamous plants on earth. They serve as food for other animals. In dead bodies they can be used to estimate the time of death with the succession of insects (forensic entomology). They receive special attention because many are harmful to crops and food.

2.2.3 Order Lepidoptera

The order Lepidoptera is one of the subdivisions of the class Insecta. Consisting of butterflies and moths, they are well known and easily recognised by the scales on their wings.

They are holometabolous, oviparous and go through the life cycle of egg, larva, pupa, imago and adult. During the larval period, they are also called caterpillars. In general, the body of caterpillars is cylindrical and made up of thirteen segments - three thoracic and ten abdominal - and a well-differentiated head, usually with six ocelli and a pair of small antennae, as well as a mouth apparatus specialised for grinding.

> Caterpillars are closely related to their host plants, and each species of caterpillar feeds on a specific species or family of plants. Therefore, the presence of a butterfly species in a given location indicates the presence of the plant species on which the caterpillars feed (host plants), and butterflies can therefore be used to indicate the presence of plant species (DA SILVA, 2008).

Each thoracic segment has a pair of legs; abdominal segments 3-6 and 10 usually have a pair of false legs, which differ in that they are more fleshy and have different segmentation and usually have tiny hooks or claws at the ends (BORROR and DELONG, 1969). The caterpillar's body may or may not have spines which, when broken when touched, communicate with certain epidermal glands in the caterpillar's skin, which release an urticating liquid and cause the skin to burn.

The pupa or chrysalis is the stage that precedes the adult form. In the pupal stage, the insect stops feeding, the caterpillar makes its last moult and shrinks until it forms the pupa or chrysalis, and the insect metamorphoses. At this stage, some insects hang from a silk thread woven by them, others weave a cocoon to protect themselves, others wrap themselves in leaves and bury themselves in the ground. When they emerge from the pupa, the butterfly or moth spreads its wings and the membrane becomes dry and it can then fly. The length of the pupal stage varies depending on the species.

Adults have wings with scales responsible for the animal's colouration. Relatively large compound eyes with a large number of facets or omatids.

Most moths have two ocelli near the edge of their eyes. The spirotromba (Figure 1) is used to suck in food, which can be nectar (nectarivorous) or decaying fruit (frugivorous).

> The nectar-eating group, which feeds on nectar, comprises the families Papilionidae, Pieridae, Lycaenidae, Hesperiidae and some subfamilies of Nymphalidae; and the frugivorous group, which feeds on fermented fruit and plant exudates, droppings and carcasses, comprising most of the nymphalids belonging to the Satyrinae, Brassolinae, Morphinae, Charaxinae, Biblidinae subfamilies and some Nymphalinae tribes (FREITAS *et al.*, 2006. *apud* SANDES and PALUCH, 2009).

When they feed, they end up playing an important role in pollination, ensuring the exchange of genetic material between species because no two organisms are the same, making them more resistant to environmental variations. Some species, such as the silkworm, have atrophied mouthparts, meaning that they don't feed as adults, but manage to subsist on the nutritional reserves they accumulate in the larval stage.

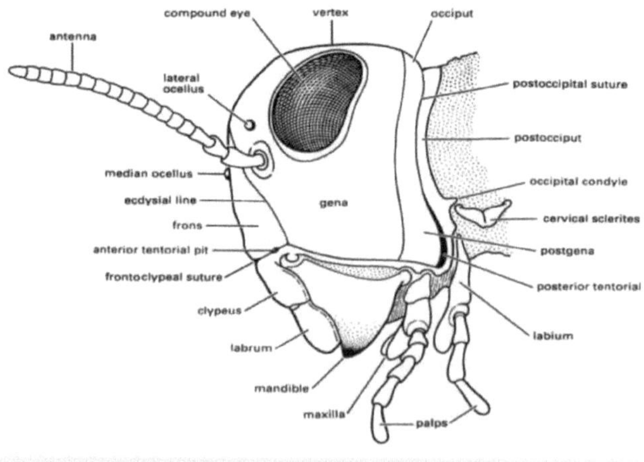

Figure 1 - External morphology of the Lepidoptera head
Source: TORRE, 2014

2.2.3.1 Classification of Lepidoptera

One of the first divisions of the order Lepidoptera into two suborders Rhopalocera (butterflies) and Heterocera (moths), based mainly on the characteristics of the antennae.

Taking venation into account, McDunnough (1938,1929) and Forbes (1923, 1948, 1954) arranged the order into two suborders Jugatae or Homoneura and Frenatae or Heteroneura (BORROR and DELONG 1969). The Jugate suborder has similar venation on the forewings and hindwings and has a yoke and no frenulum. The Frenatae have a frenulum, no yoke and reduced venation on the hind wings.

Sequentially, the order was arranged according to the average size of the insects as microlepidoptera and macrolepidoptera.

> The most recent arrangements separate the Micropterigidae into their own suborder and the Zeugloptera (some consider them to be an order related to the Lepidoptera and Trichoptera), with the other Lepidoptera divided into two suborders, iMonotrysia and Ditrysia, based mainly on the number of female genital openings (TRIPLEHORN and JOHNSON, 2011).

The *checklist by* Hodges *et al.* (1983), presented by Triplehorn and Johnson, (2011) shows the division of the order into families and superfamilies.

In the revision of Kristensen *et al.* (2007) presented by Rafael (2012), the higher categories of superfamilies (Figure 2) are abandoned, which constituted an arbitrary hierarchisation, where it was redundant to use names for suborders with a superfamily and a family, such as Zeugloptera, Aglossata and Heterobathmiina and Heterobathmioidea. Glossata, Coelolepida and Myoglossata are referred to simply as clades. Many arrangements have been made, however the newest are experimental classifications.

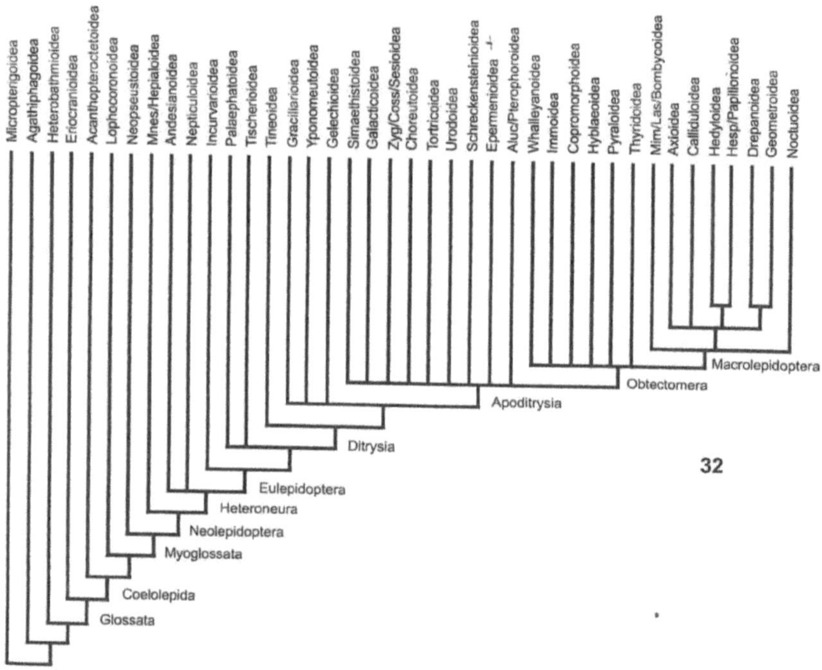

Figure 2 - Phylogeny of the current Lepidoptera Superfamilies
Source: RAFAEL, 2012

2.2.3.2 Characters used in the classification of Lepidoptera

The main characters used to identify adult lepidopterans are venation, shape, coupling mode and wing scales. Also important are the antennae, mouthparts, the presence or absence of ocelli, legs, genitalia, abdomen and general characteristics such as size and colour (TRIPLEHORN and JOHNSON, 2011).

2.2.3.3 Wings

The wings on each side operate together through a set of basic mechanisms: fibula, yoke, frenulum and humeral angle. The fibula (Figure 2B, *fib*) is a small lobe at the base of the forewings, which overlaps the base of the hindwings.

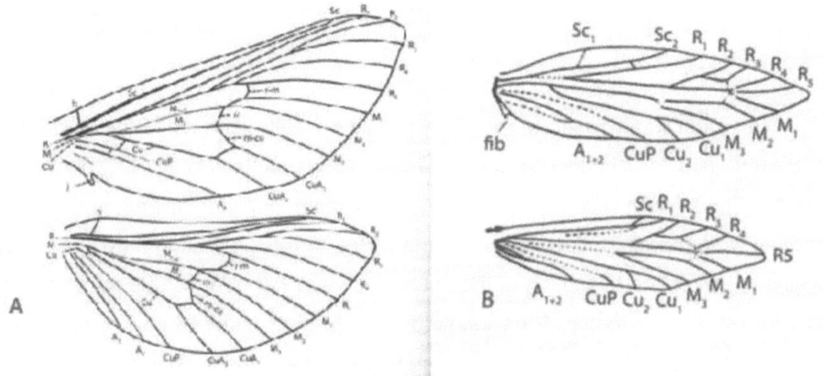

Figure 3 - A, homoneura venation of *Sthenopis* (Hepialiade), *j*, yoke; B, *Micropterys* (Micropterigidae), *fib*, fibula

Source: TRIPLEHORN and JOHNSON, 2011

The yoke (Figure 2A, *j*) is a small digitiform lobe at the base of the forewings that overlaps the base of the anterior edge of the hindwings. Frenulum is a bristle (males) or a group of bristles (females) originating from the humeral angle of the hind wing (Figure 2, *f*). Humeral angle is the anterior basal portion of the wings. Most lepidopterans have more or less triangular forewings and rounded hindwings (TRIPLEHORN and JOHNSON, 2011). The wings are reinforced by a series of sclerotised veins, and the pattern of wing venation is used for taxonomy as a means of identification.

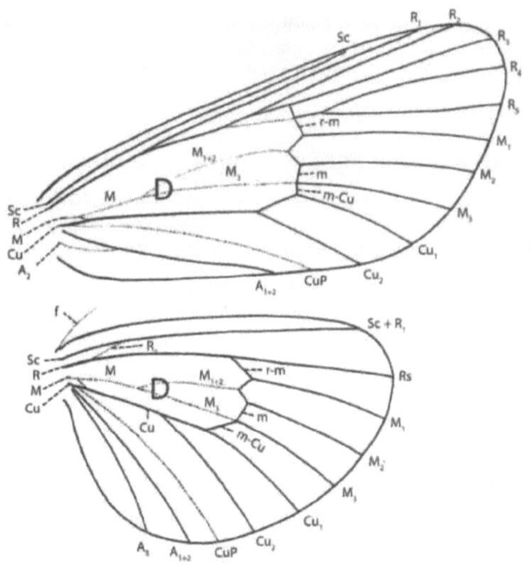

Figure 4 - Generalised heteroneural venation. The veins represented by dotted lines are atrophied or lost in some groups. *D*, disc cell; *f*, frenulum

Source: TRIPLEHORN and JOHNSON, 2011

1.1.1.1.1 Alar venation

The most widely used terminology based on wing venation is the Comstock-Needham system (Comstock and Needham 1898, 1899) explained by Triplehorn and Johnson, 2011. The system recognises a series of six main longitudinal wing veins: costa (C) on the main wing margin, followed by subcosta (Sc), radius (R), median (M), ulna (Cu) and anal veins (A). With the exception of the costa, the veins can branch; the subcosta can branch only once. The branches of the longitudinal veins are numbered from anterior to posterior with subscript numerals. The transverse veins connect the main longitudinal veins and are named accordingly. The spaces in the alar membrane are called cells, which can be open or closed.

According to Comstock's system cited by Triplehorn and Johnson, 2011, there are two types of venation in the order Lepidoptera, homoneura and heteroneura venation.

Homoneura is the similar venation on the forewings and hindwings, with the same number of R-branches on both (Figure 2A). It occurs in the superfamilies Micropterigoidea, Eriocranioidea, Acanthopteroctetoidea and Hepialioidea. They have a single or two-branched subcostal vein, a five-branched or occasionally six-branched radial vein, a three-branched median vein and three anal veins.

Heteroneura the venation of the hind wings is reduced and Rs is not branched (Figure 3), occurring in the other subfamilies. The radial of the forewings has five branches or occasionally fewer, in the hindwings the radial sector is not branched and R1 merges with the subcosta. The basal portion of the median is often atrophied, causing the discal cell to occupy the central part of the wing. The first anal vein is often atrophied or fused with A2. The subcosta of the forewings is almost always free in relation to the discal cell and is situated between it and the costa. The radial branches originate at the anterior margin of the disc cell or up to its upper edge. The radial branches are often pedunculated (fused) for some distance beyond the disc cell. There are cases where some branches of the radial fuse again beyond their separation points to form accessory cells (Figure 4, *acc*).

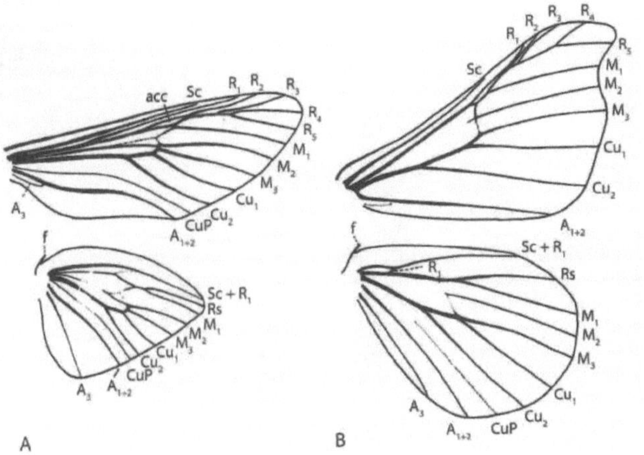

Figure 5 - A, wings of *Prionoxystus* (Cossidae); B, wings of *Bombyx*) Bombycidae. *acc*, accessory cell; *f*, frenulum

Source: TRIPLEHORN and JOHNSON, 2011

The three branches of the median originate at the apex of the disc cell in both wings. The point of origin of M2 is an important feature for distinguishing different groups: when it originates halfway along the outer margin of the disc cell, or anterior to the midpoint, the cubital vein appears to have three branches; when M2 originates closer to M1, the cubital vein appears to have four branches.

In heteroneurons the variation in venation mainly involves the fusion of Sc + R1 and the number of anal veins. In some cases R is separated from Sc at the base of the wing and R1 appears as a vein between Rs and Sc and somewhere along the upper margin of the discal cell (Figure 3B). R1 also merges with Sc and, judging by the pupal tracheation, the vein that reaches the wing margin is Sc - the trachea of R1 is always small, this vein at the margin is called Sc + R1. Sc and R in many cases are fused at the base or can be separated at the base and fused for a short distance along the upper margin of the disc cell. Most authors call the 1ª of Comstock (1940) the CuP; 2A the A_{1+2} and 3A the A_3 .

2.2.3.3 Aerials

Lepidoptera have elongated antennae that vary in shape. Based on this sensory organ, a quick distinction can be made between butterflies and moths. Butterfly antennae are slender and swollen at the apex, while moth antennae are filiform, setaceous or feathery (Figure 6).

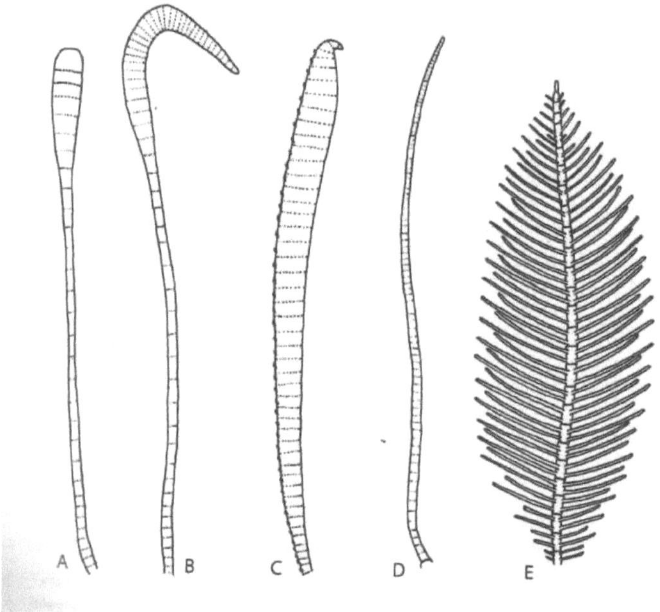

Figure 6 - Lepidoptera antennae. A, *Colias* (Pieridae); B, *Epargyreus* (Hesperidae); C, *Hemaris* (Sphingidae); D, *Drasteria* (Noctuidae); E, *Callasamia* (Saturnidae).
Source: TRIPLEHORN and JOHNSON, 2011

2.2.3.4 Legs

The characteristics of the legs used for identification are simple or bifid tibial and tarsal claw spurs, the presence or absence of spines on the legs and sometimes the structure of the epiphysis (a mobile spur-like structure on the inner surface of the forelegs). In some butterflies the forelegs are very short, particularly in the Nymphalidae (Danaidae, Satyridae and Libytheinae) and in the males of the Riodininae.

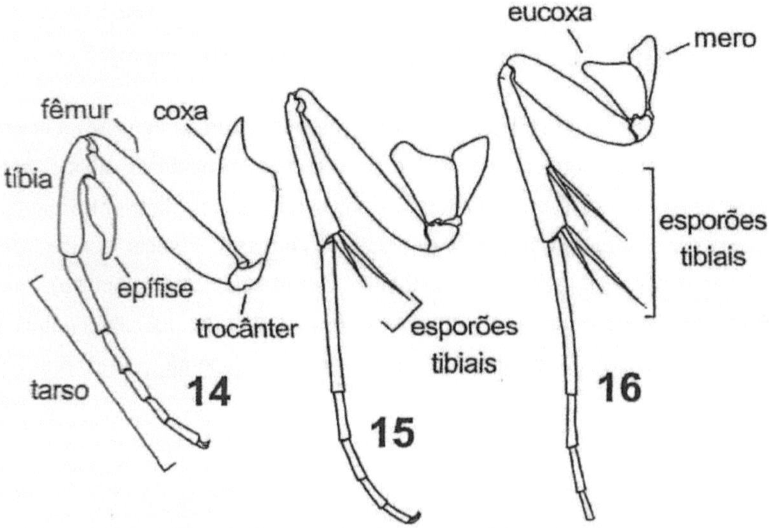

Figure 7 - External morphology of the legs
Source: RAFAEL, 2012

2.3 POLLINISATION

Fossil records show that the first flowers contain characteristics that indicate pollination by insects: small anthers, little *pollen* production, the presence of *pollenkitt* in the grain (which helps it to stick to the pollinator) and the size of the pollen grain unsuitable for anemophily (ROBERTSON, 1917, GRIMALDI; ENGEL, 2005 *apud* NETO, 2009). Insect pollination probably drove the evolution and diversification of angiosperms (LIMA, 2000).

Pollination is the transfer of pollen grains from the anthers of a flower to the stigma (part of the female reproductive system) of the same flower or another flower of the same species (EMBRAPA, 2013). The pollen grain must fertilise the ovule for the fruit to form. It is important for genetic material to be exchanged between species because no two organisms are the same, making them more resistant to environmental variations.

> The more attractive the plants, the more diligent the animal visits and the greater the chance of successful seed production and cross-pollination, which increases genetic variability and, consequently, the potential for adaptation to the vicissitudes of the environment (RAVEN, 2007).

To attract biotic agents, plants have various attractants, such as nectar, pollen and oils, which are used to feed and reproduce these animals. Signalling floral resources is a way of warning and guiding biotic agents to the presence of food. These include eye-catching colours and smells, each signal is adapted according to the perception of the agent, and all these characteristics are called pollination syndromes (NETO, 2009).

Pollen transfer can occur in the same flower - self-pollination, or in different flowers - cross-pollination. The latter can occur with the participation of biotic or abiotic factors. Among the syndromes in which biotic factors are involved, we can highlight ornithophily (birds), chiropterophily (bats) and entomophily (insects in general); insects include beetles (cantophilia), flies (myophilia), bees (melittophilia), butterflies (psychophilia), moths (phallenophilia); those that depend on abiotic factors include hydrophilia (water) and anemophilia (wind). Species that depend on animals to disperse pollen are called zoophytes.

2.3.1 Psychophilia

Psychophily is the name given to pollination by butterflies. Flowers pollinated by butterflies have certain characteristics such as: daytime anthesis, tubular corollas, showy flowers and little odour. Species adapted to pollination have a specialised mouthpart: a spirotromba, which can be inserted into the tubular corollas to suck nectar. The pollen grains attach themselves to the hairs and scales, which are then transported to the other flower.

2.4 PERMANENT PRESERVATION AREAS - APPS

A Permanent Preservation Area (APP), created by environmental law, aims to preserve water resources, landscape, geological stability, biodiversity, protect the soil and ensure the well-being of populations (BRASIL, Federal Law 12.651/2012). According to Mota (1995), although they are not a measure of total efficiency, they represent a valid measure for preserving surface water resources, with the following advantages:

- Sanitary protection for reservoirs and watercourses, preventing surface and subsurface access for pollutants;
- Rainwater drainage, protecting adjacent areas from flooding;
- Preservation and promotion of vegetation on the banks of water resources, guaranteeing the protection of typical fauna and flora. The shade provided by the vegetation also helps to maintain the water temperature suitable for aquatic fauna;
- Preventive action against erosion and the consequent silting up of water bodies;
- Landscape and ecological conservation.

CHAPTER 3

METHODOLOGY

3.1 RESEARCH METHODOLOGY

This research was carried out through bibliographical research, field research and laboratory research, based on the collection of specimens in the APPs, assembly and determination in the laboratory, with the help of data available in specialised literature.

3.2 RESEARCH LOCATION

Domain strip of the BR-116 motorway, from Mafra to Papanduva, SC.

3.3 RESEARCH DELIMITATION

3.3.1 Universe

Permanent Preservation Areas (APPs) on the marginal strip of the BR-116 motorway, stretching from Mafra, SC (km 015 + 400) to Papanduva, SC (km 058 + 770), represented by its water bodies, flora and fauna.

3.3.2 Sample

Diverse specimens of butterfly entomofauna (Lepidoptera) collected in the study area.

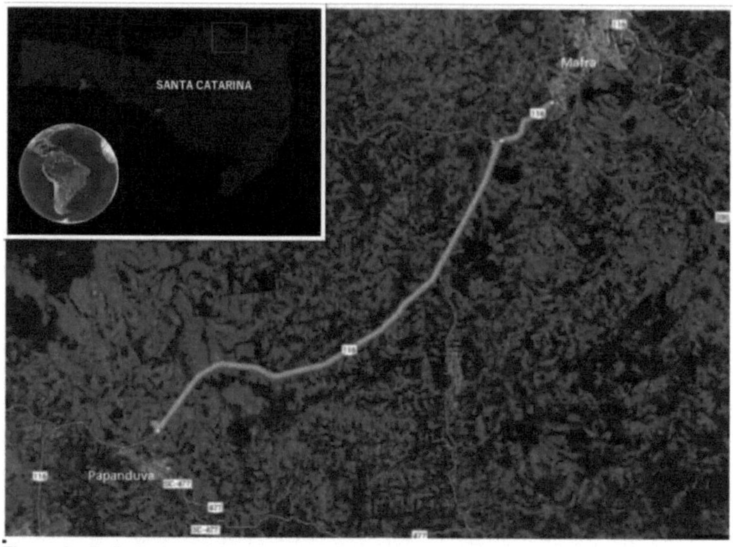

Figure 8 - Delimitation of the research site

SOURCE: Google Maps, 2014

3.3.3 Inclusion Criteria

Only the butterfly species (Lepidoptera) found in the study area were included in the survey.

3.3.4 Exclusion Criteria

Insects from other taxonomic groups and butterflies found outside the Permanent Preservation Area (APP) were not included in the survey.

3.3.5 Criteria for terminating or suspending research

The research ended in the first fortnight of October 2014, after the CBT report had been drawn up.

3.3.6 Critical Analysis of Risks and Benefits

This research posed risks to the researcher, such as sunburn, falls, being run over and snakes. To mitigate the risks, sunscreen was used, as well as appropriate footwear for the field, PPE and the recommendations of the Concessionaire's safety team.

The benefits were to add to the Concessionaire's knowledge about the state of preservation of the Permanent Preservation Areas (PPAs) in the motorway's right of way and to highlight the importance of pollination processes in the recovery of these areas.

3.4 METHODOLOGICAL PROCEDURES

3.4.1 Research instruments

The research was carried out using the following instruments and equipment: field notebook, photographic camera, entomological net, killing bottle, butter paper envelopes for packaging and transporting the specimens collected, entomological pins, tweezers, Lepidoptera extenders, dehydration oven, entomological drawers, stereomicroscope, microcomputer, printer, collection labels and identification of the specimens.

3.4.2 Bibliographical Research Procedures

The research was carried out using scientific articles, books, manuals and magazines related to the proposed topic.

3.4.3 Field Research Procedures

The field research was carried out in the following stages:

- Delimitation of permanent preservation areas - APP.

- Catching and killing insects.
- Packaging the insects in envelopes.
- Note the collection site.

3.4.4 Laboratory Research Procedures

The laboratory research was carried out in the following stages:

- Deposit the insects in a humid chamber.
- Mounting the insects on Lepidoptera stretchers.
- Take the insects assembled on the stretcher to the greenhouse.
- Remove the insects from the greenhouse.
- Dismantle the stretcher bars;
- Determine the family of each insect collected;
- Store insects in entomological boxes.

3.5 DESCRIPTION OF THE FIELD RESEARCH

Eight field trips were made, on random dates depending on the weather, between March 2014 and September 2014, with a sampling effort of 27 hours. On the days chosen for the collections, the weather was sunny and with little wind, and the collections took place between 12pm and 5pm.

During the outings, the insects were captured using an entomological net and killed using a killing chamber mounted in a glass jar with a layer of absorbent paper, a layer of plaster soaked in ethyl acetate ($C_4H_8O_2$); once dead, the insect was packed in triangular envelopes made of butcher paper (figure 12) and duly identified.

Figure 9 - Collecting butterflies
SOURCE: SCHELBAUER, 2014

Figure 10 - Removing the insects from the entomological net
SOURCE: SCHELBAUER, 2014

Figure 11 - Slaughter of Lepidoptera specimens using a killing bottle
SOURCE: SCHELBAUER, 2014

Figure 12 - Storing butterflies in envelopes
SOURCE: SCHELBAUER, 2014

After each collection, the insects are stored in tightly sealed plastic bags in the fridge to prevent microorganisms from breeding until they are assembled.

A total of 138 insects were collected from 16 Permanent Preservation Areas (APPs).

Table 1 - Insects collected by Permanent Preservation Area - APP

Municipality	Km	Number of collections	Number of insects collected
Mafra	015 + 400	3	19
Itaiopolis	020 + 950	3	16
Itaiopolis	021 +100	1	7
Itaiopolis	021 +190	1	5
Itaiopolis	021 + 435	2	14
Itaiopolis	021 + 450	1	7
Itaiopolis	026 + 780	1	2
Itaiopolis	026 + 950	3	16
Itaiopolis	029 +810	1	9
Itaiopolis	031 +815	2	12
Itaiopolis	037 + 755	1	2
Papanduva	046 + 570	1	4
Papanduva	058 + 265	2	10
Papanduva	058 + 770	1	8
Papanduva	061 + 900	1	3
Papanduva	061 + 925	1	4

SOURCE: SCHELBAUER, 2014

3.6 DESCRIPTION OF LABORATORY RESEARCH

After all the specimens had been collected, they were mounted. Insects that had been dead for more than two days and were sclerotised were placed in a humid chamber for 24 hours so that the structures became malleable. After being removed from the humidity chamber, the insect was fixed to an entomological pin (Figure 13;A) appropriate to its size and placed in the centre of the stretcher (Figure 6;H), securing the wings with strips of butcher paper (Figure 13;E) and map pins (Figure 13;D), positioning the posterior edge of the front wing at a 90° angle to the insect's body; the antennae and abdomen are positioned with the aid of pins (Figure 13;B). Tweezers (Figure 13;F) are used to attach the entomological pin to the insect and the insect to the spreader, as well as to open the wings, adjust the legs and antennae, avoiding damage caused by direct contact with the fingers.

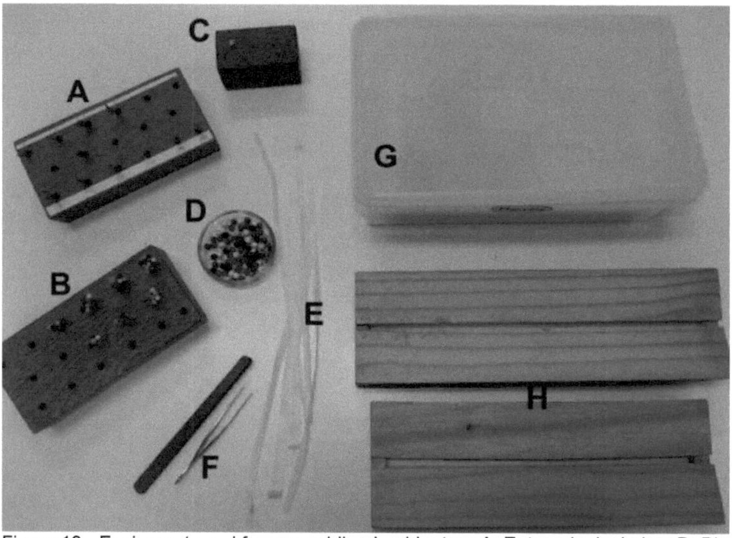

Figure 13 - Equipment used for assembling Lepidoptera A. Entomological pins; B. Pins for positioning the insects; C. Measures for positioning the insect; D. Pins for attaching the wings; E. Strips for attaching the wings; F. Tweezers; G. Humid chamber; H - Lepidoptera stretchers
Source: SCHELBAUER, 2014

Figure 14 - Mounting the specimen on the Lepidoptera stretcher
SOURCE: SCHELBAUER, 2014

After assembly, the stretcher bars are placed in the oven for approximately 24 hours at a temperature of 35°C.

Figure 15 - Greenhouse with fountains
SOURCE: SCHELBAUER, 2014

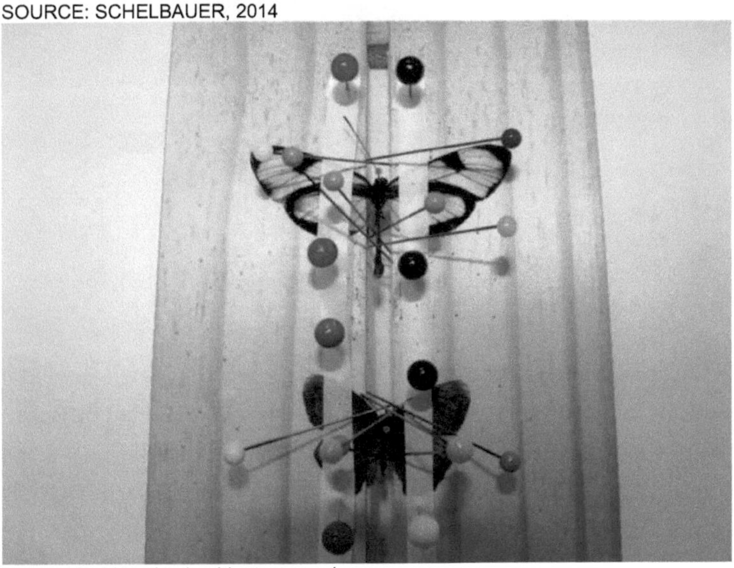

Figure 16 - Mounting Lepidoptera specimens
SOURCE: SCHELBAUER, 2014

After drying in the oven, the appendix was carefully removed and disassembled so that it wouldn't break; first the pins were removed from the wings and abdomen, then the strips of parchment paper from the wings and the other pins. If the appendix broke, it was glued back on with colourless nail polish.

The assembled and dehydrated insects were labelled with the collection data:

Country, State, Municipality, Kilometre, Date of Collection, Name of Collector. The insects are stored in entomological boxes until they are determined (figure 8).

BRASIL, SC. ITAIÓPOLIS
01.III.2014
BR – 116, km 021 + 450
K. SCHELBAUER, col.

Figure 17 - Label model: 2 x 1 cm
SOURCE: SCHELBAUER, 2014

Figure 18 - Packaging the insects in an entomological box
SOURCE: SCHELBAUER, 2014

Once the specimens had been prepared, the researcher began to determine them using artificial keys specific to Lepidoptera. The well-established keys presented by Borror and DeLong (1969) were used in preference to other alternatives considered experimental (TRIPLEHORN and JOHNSON, 2011). The structures were examined under a QUIMIS stereoscopic microscope, model Q714Z2.

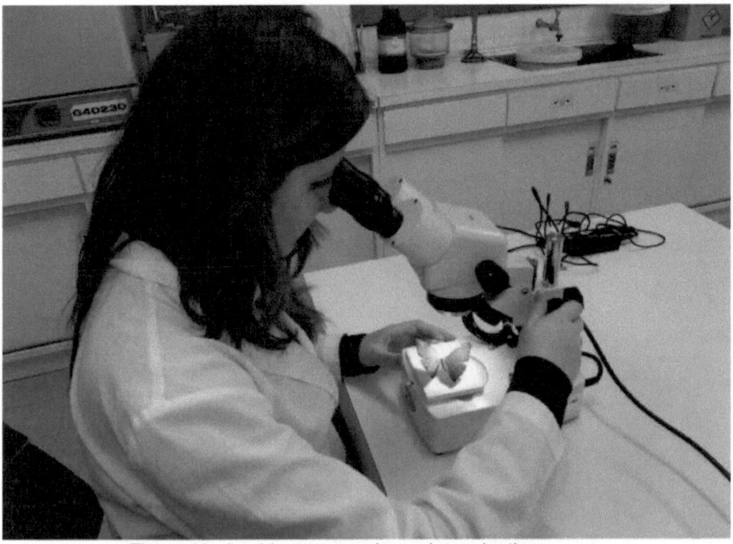

Figure 19 - Lepidoptera specimen determination
SOURCE: SCHELBAUER, 2014

CHAPTER 4

ANALYSING AND INTERPRETING THE RESULTS

Eight field expeditions were carried out between March and September 2014 in the APPs along the BR-116 motorway, from Mafra to Papanduva, SC, and 138 individuals from nine Lepidoptera families were collected (Graph 1).

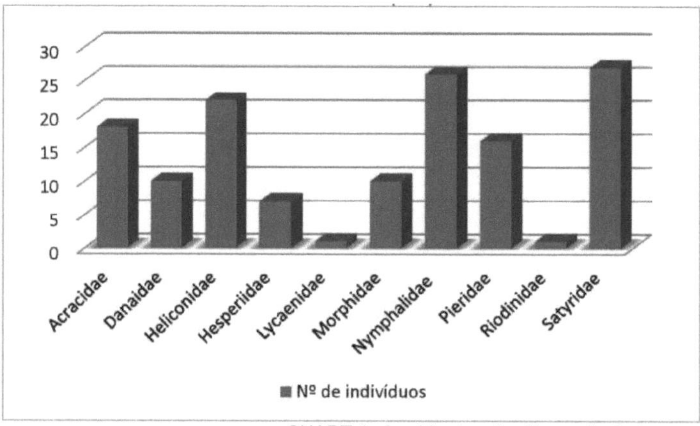

CHART 1 - Lepidoptera families collected
Source: SCHELBAUER, 2014

The most representative family was Nymphalidae, with 26 specimens collected, followed by Heliconidae with 22 specimens. The families with the fewest specimens collected were Lycaenidae and Riodinidae, with only one specimen each.

Family Description

The following families have been described on the basis of the arrangement by McDunnough (1938, 1939) and Forbes (1923, 1948, 1954) with some modifications presented by Borror and DeLong (1969).

Family Nymphalidae - very reduced forelegs, only the middle and hind legs are used for walking, no claws. The discal cell can be open or imperfectly closed. Radial vein on the forewing with 5 branches.

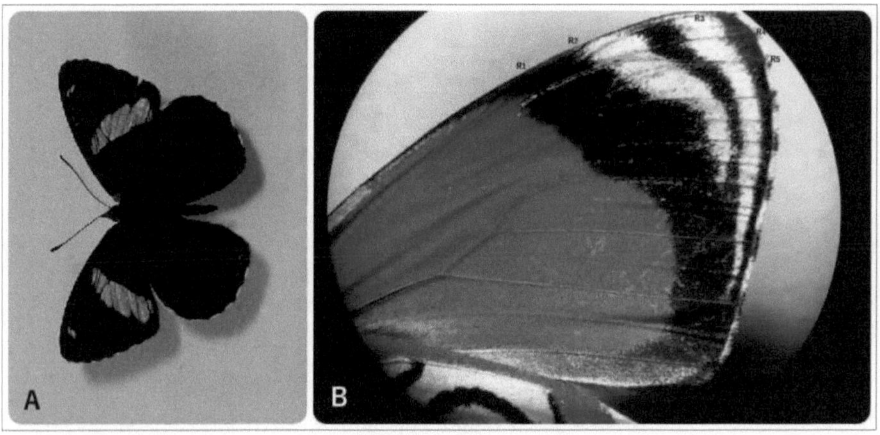

Figure 20 - Nymphalidae
A. General aspect; B. Wing with 5 branches of the radial vein SOURCE: SCHELBAUER, 2014

Family Hesperiidae - known as little devils because of their fast and erratic flight. They are mostly small and robust-bodied. The antennae are separated at the base and recurved at the apex or hooked; none of the five branches of R on the forewings are pedunculated and all originate from the discal cell.

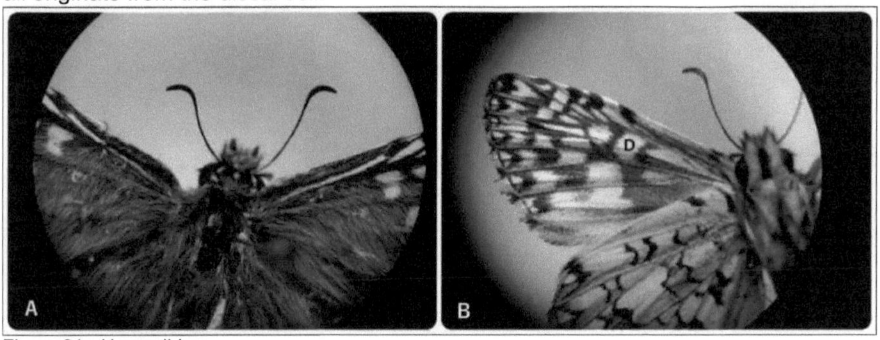

Figure 21 - Hesperiidae
A. Antennae separated at the base and hook-shaped; B. Branches of R originate in the discal cell (D)
SOURCE: SCHELBAUER, 2014

Family Pieridae - white, buds. They are medium to small, white or yellowish in colour with black spots on the basal margin of the wings. The forelegs are well developed and the tarsal claws are bifid. The radial vein on the forewing usually has three or four branches (rarely five).

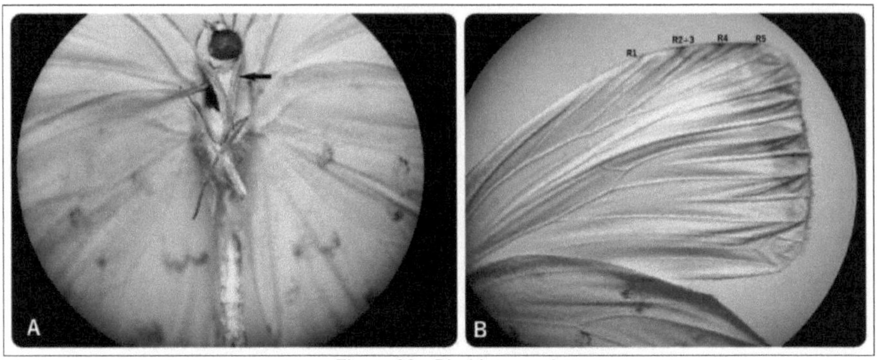

Figure 22 - Pieridae specimen
A. Normal forelegs (arrow); B. Branches of vein R SOURCE: SCHELBAUER, 2014

Family Heliconiidae - narrow, elongated wings. The humeral vein of the hind wings curves towards the base of the wing. Wings densely covered with brightly coloured opaque scales.

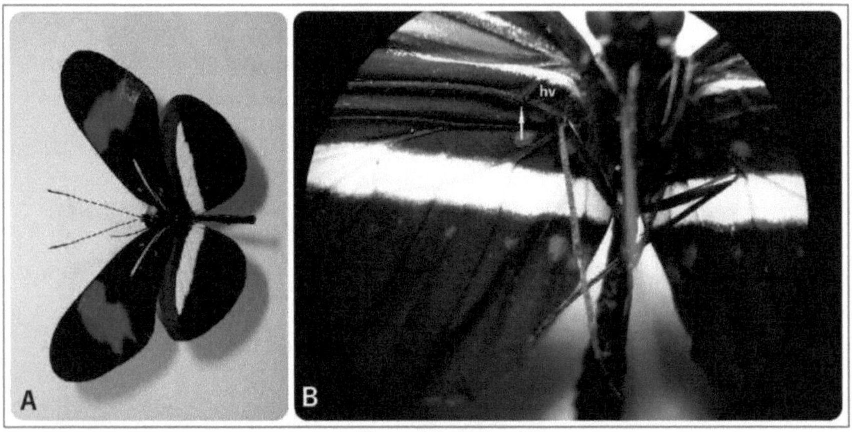

Figure 23 - Heliconiidae
A. General appearance; B. Humeral vein (hv) curving towards the base of the wing (arrow) SOURCE: SCHELBAUER, 2014

Family Acracidae - have a wingspan of between 5 and 6 cm; humeral vein curving towards the wing tip, and closed disc cells on both wings.

Figure 24 - Acracidae

A. General appearance; B. Discal cell (D) closed on both wings, humeral vein (hv) curving towards the wing tip (arrow).

SOURCE: SCHELBAUER, 2014

Family Danaidae - usually brightly coloured. Small forelegs without claws, not used for locomotion. R forewing with five branches; closed discal cell. Small third anal vein (3A) on forewings.

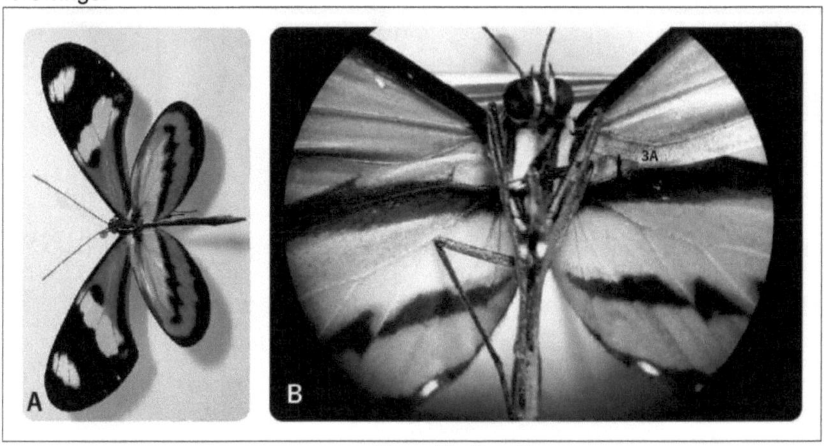

Figure 25 - Danaidae

A. General aspect; B. 3A of the anterior wing present (arrow) SOURCE: SCHELBAUER, 2014

Family Lycaenidae - small, brightly coloured butterflies. Slender body and white-striped antennae, they also have a line of white scales around the eyes. The forewings have an R with three or four branches and the M1 originates at or near the apex. Hind wings without humeral vein.

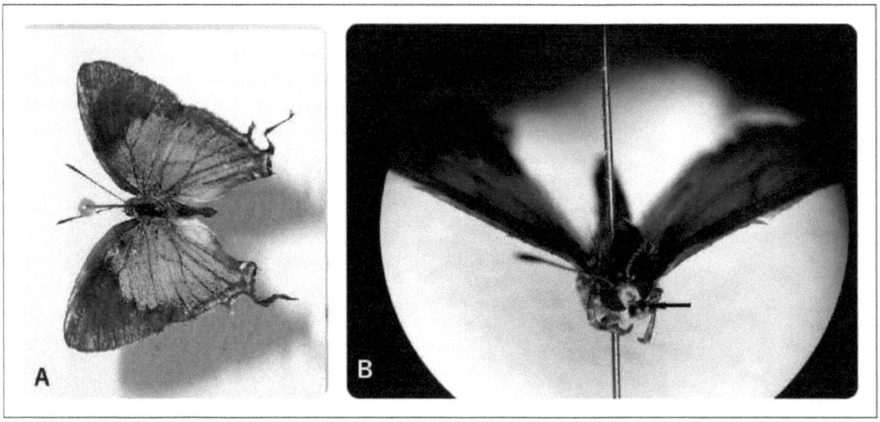

Figure 26 - Lycaenidae
A. General appearance; B. Line of white scales around the eyes (arrow) SOURCE: SCHELBAUER, 2014

Family Satyridae - medium to small, greyish or brown in colour with ocellar spots on the wings. Some ribs of the forewing are quite dilated at the base; antennal dilation gradually widening. Closed discal cell on the hind wings.

Figure 27 - Satyridae
A. General appearance; B. Forewing veins greatly dilated at the base (arrow); C. Ocellar spots SOURCE: SCHELBAUER, 2014

Family Morphoidae - brightly coloured, sometimes metallic with a predominance of blue. The discal cell is open on the hind wings and distinctly closed on the forewings.

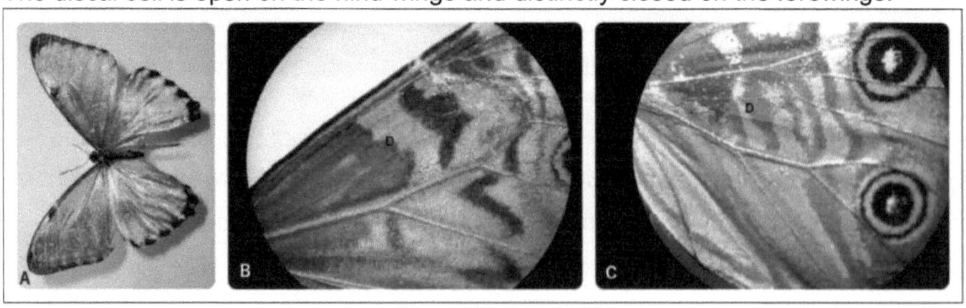

Figure 28 - Morphoidae
A. General appearance; B. Anterior wing with closed disc cell (D); C. Posterior wing with closed disc cell (D)

SOURCE: SCHELBAUER, 2014

Family Riodinidae - dark-coloured butterflies; hind wings with a thickened costal vein in the region of the humeral angle; Sc with a basal thorn (humeral vein).

Figure 29 - Rhodinidae

A. General appearance; B. Posterior wing C thickening towards the humeral angle and Sc with basal spine
(humeral vein).

Source: SCHELBAUER, 2014

TABLE 1 - Families and number of individuals collected per APP

	015 + 400	020 + 950	021 + 100	021 + 190	021 + 435	021 + 450	026 + 780	026 + 950	029 + 810	031 + 815	037 + 755	046 + 570	058 + 265	058 + 770	061 + 900	061 + 950
Acracidae	8				2	1		1	1	2			3			
Danaidae		1	2	1	1					2			1	1	1	
Hesperiidae	2							2			1	1	2			
Heliconidae	2	2	1	1	2	2		5	1	2				3		
Lycaenidae													1			
Morphoidae			3	2	1			1		1		1		1		
Nymphalidae	3	4		1	5	2			7	1			2	1		
Pieridae	4	5				1		1		2		2			1	
Riodinidae														1		
Satyridae		4	1		3	1	2	6		2	1		1	1	1	4
Total number of families	**5**	**5**	**4**	**4**	**6**	**5**	**1**	**6**	**3**	**7**	**2**	**3**	**6**	**6**	**3**	**1**
Total number of	**19**	**16**	**7**	**5**	**14**	**7**	**2**	**16**	**9**	**12**	**2**	**4**	**10**	**8**	**3**	**4**

individuals																

Source: SCHELBAUER, 2014

Analysing Table 1, it can be seen that the Permanent Preservation Area (PPA) with the greatest variety of families was at km 031 + 815, with 7 families, followed by the PPAs at km 021 + 435, km 026 + 950, km 058 + 265 and km 058 + 770, with 6 families each. The PPAs with the least variety of families were those at km 026 + 780 and km 061 + 950; however, only one collection was made in each of the latter two, as shown in Table 1.

CHAPTER 5

FINAL CONSIDERATIONS

Eight field expeditions were carried out between March and September 2014 in the APPs along the BR-116 motorway, from Mafra to Papanduva, SC, and 138 individuals from ten Lepidoptera families were collected.

The APP at km 031 + 815 showed the greatest diversity of Families (Acracidae, Danaidae, Heliconidae, Morphoidae, Nymphalidae, Pieridae, Satyridae).

The APPs at km 026 + 780 and km 061 + 925 showed less variety, with all the specimens belonging to the Satyridae family, possibly because only one expedition was carried out in each of these APPs.

It was possible to record the occurrence of 10 families of Lepidoptera in the stretch: Acracidae, Danaidae, Hesperiidae, Heliconidae, Lycaenidae, Morphoidae, Nymphalidae, Pieridae, Riodinidae and Satyridae. The most representative family was Nymphalidae.

It should be emphasised that the greater the diversity of individuals and the larger their population, the greater the transport of pollen between forest fragments, which increases the likelihood of flowers fertilising and the consequent production of fruit and seeds, desirable aspects for environmental restoration.

CHAPTER 6

REFERENCES

BORROR, D.J. & DELONG, D.M. **Study of Insects.** São Paulo, Edgard Blucher, 1969.

BRAZIL. Constitution (2012). Law No. 12.651, of 25 May 2012. **Provides for the Protection of Native Vegetation**: Federal Law. Available at: <www.planalto.gov.br/ccivil_03/_Ato2011/2014/2012/Lei/L12651.htn>. Accessed on: 12 March 2014.

DA SILVA, Gabriela Corso. 2008. **DIVERSITY OF NYMPHALIDAE BUTTERFLIES IN THE ATLANTIC FOREST OF THE LAGOA DO PERI MUNICIPAL PARK, FLORIANÓPOLIS, SC.** Available at: <http://www.ccb.ufsc.br/biologia/TCC-BIOLOGIA-UFSC/TCCGabrielaCorsodaSilvaBioUFSC-08-2.pdf> Accessed on: 13 Apr. 2014.

BRAZILIAN AGRICULTURAL RESEARCH COMPANY - EMBRAPA. **Pollination.** Available at: <http://www.cpamn.embrapa.br/apicultura/polinizacao.php> Accessed on: 13 Sep 2014.

GOOGLE MAPS. **Mafra to Papanduva.** 2014. Available at: https://www.google.com.br/maps Accessed on: 02 October 2014.

HICKMAN JUNIOR, Cleveland P.; ROBERTS, Larry S.; LARSON, Allan. **Integrated principles of zoology.** 11.ed. Rio de Janeiro: Guanabara Koogan, 2009.

LIMA, Cintia. **Flowers and insects: the origin of entomophily and the success of angiosperms.** Available at: <http://repositorio.uniceub.br/bitstream/123456789/2386/2/9508967.pdf> Accessed on: 18 Sep 2014.

MOTA, Suetônio. **Preservation and conservation of water resources**. Rio de Janeiro: ABES, 1995
Neto, Laércio Peixoto do Amaral. **POLLINISATION BIOLOGY: INTERACTIONS BETWEEN BEEES (Hym., Apoidea) AND FLOWERS WITH INVERTED KILLS IN FABACEAE.** 2009. Available at: http://www.iap.pr.gov.br/arquivos/File/Pesquisa%20em%20UCs/Projetos%20de%20Pesquisas%20Autorizados%20em%202009/INTERACOES_ENTRE_ABELHAS_E_FABACEAE.pdf> Accessed on: 22 Sep 2014.

RAFAEL, J.A.; MELO, G.A.R.; CARVALHO, C.J.B.; CASARI, S.A.; CONSTANTINO, R. (Orgs.). **Insects of Brazil: Diversity and Taxonomy.** Holos, Editora, Ribeirão Preto, 2012.

RAVEN, P.H.; EVERT, R.F. & EICHHORN, S.E. **Plant Biology**. 7ª . ed. Coord. Trad. J.E.Kraus. Editora Guanabara Koogan, Rio de Janeiro. 2007.

SANDES, Luiz Cláudio Mattos; PALUCH, Márlon. **Preliminary list of the butterflies (Papilionoidea and Hesperioidea) of Dois Irmãos State Park, Recife, PE.**

Available at:

http://www.contabeis.ufpe.br/propesq/images/conic/2009/anais%20(E)/conic/pibic/20/092041139SCPP.pdf> Accessed on: 13 Mar. 2014.

TRIPLEHORN, C.A. & JOHNSON, N.F. **Study of Insects.** 7th Edition. Cengage Learning. São Paulo. 2011.

TORRE, Rafael. **External morphology of insects general entomology.** Available at: <http://slideplayer.com.br/slide/1818216/> Accessed on: 19 Oct. 2014

ANNEXES

ANNEX 1 - DECLARATION

Habilitação | Grau

Português e Literaturas

de Língua Portuguesa 1º e 2º

Inglês e Literaturas

de Língua Inglesa 1º e 2º

Curitiba, 14 / 12 / 1978

CARLOS CECY

Delegado Regional

MINISTÉRIO DA EDUCAÇÃO E CULTURA

SECRETARIA DE APOIO ADMINISTRATIVO

Delegacia Regional (DR- 9)

Certificado de Registro de Professor

REGISTRO "L" – Nº 267.650

NOME Elenir de Fátima Ruthes

CURSO Letras: Potuguês-Inglês

ESCOLA Fac.de C.e Letras de Mafra-SC

PROCESSO Nº 2387-DR.9-78

MIX
Papier aus verantwortungsvollen Quellen
Paper from responsible sources
FSC® C105338

Printed by Books on Demand GmbH, Norderstedt / Germany